THE MILITARY EXPERIENCE.

Special Operations: TRAINING

THE MILITARY EXPERIENCE.

Special Operations: TRAINING

DON NARDO

GREENSBORO, NORTH CAROLINA

The Military Experience.
Special Operations: Training

For more information write:
Morgan Reynolds Publishing, Inc.
620 South Elm Street, Suite 387
Greensboro, NC 27406 USA

Library of Congress Cataloging-in-Publication Data

Nardo, Don, 1947-
Special operations : training / by Don Nardo.
p. cm. -- (The military experience)
Includes bibliographical references and index.
ISBN 978-1-59935-356-2 -- ISBN 978-1-59935-357-9 (e-book) 1.
Commando
troops--Training of--United States. 2. Special forces (Military
science)--United States 3. Special operations (Military
science)--United
States. I. Title.
UA34.S64N48 2013
356'.16--dc23
2012016875

Printed in the United States of America
First Edition

Book cover and interior designed by:
Ed Morgan, navyblue design studio
Greensboro, NC

Table of Contents

U.S. Navy SEAL trainees scan a room for possible threats as part of a SEAL Qualification Training exercise. Students spend two weeks learning the basics of securing a room from possible threats.

Four Navy SEAL trainees practice a room breaching exercise as part of SEAL Qualification Training instruction.

TRAINING to KILL bin Laden

"Today we should all be proud," U.S. naval officer Edward Winters said in May 2011. "That handful of courageous men, of strong will and character, have changed the course of history." Winters was talking about a small group of Navy SEALs. On May 2, 2011, they killed Osama bin Laden. He had long been the leader of the terrorist group al-Qaeda.

Since the 1990s, al-Qaeda had frequently attacked the U.S. and other Western targets. Bin Laden's and al-Qaeda's most infamous deed had been the 9/11 attacks. On September 11, 2001, several of bin Laden's followers had hijacked American airliners. They had crashed two of the planes into New York's World Trade Center towers, and soon afterward those buildings had collapsed. Another stolen plane had hit the Pentagon, in Washington, D.C. Almost 3,000 Americans had lost their lives that day.

An aerial view of the Pentagon in Arlington, Virginia, showing emergency crews responding to the destruction caused when a hijacked commercial jetliner crashed into the southwest corner of the building during the 9/11 terrorist attacks in 2001

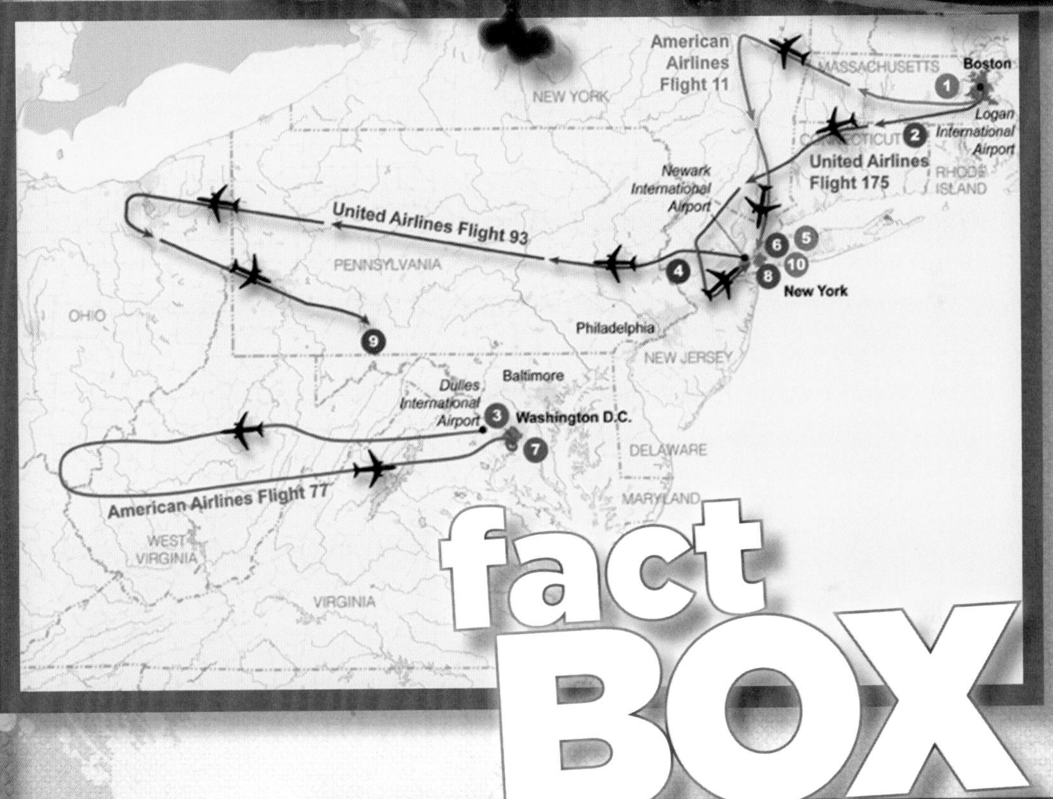

top: A United Airlines Boeing 757-222, similar to the plane that crashed in Pennsylvania

bottom: The flight paths of the four hijacked planes used in the terrorist attacks on September 11, 2001

fact BOX

Fate of the Fourth Plane

On 9/11, a fourth plane was hijacked. It appears the terrorists planned to fly it into either the White House or the U.S. Capitol, but the plane's passengers bravely moved against the hijackers. During the struggle, someone lost control. And the plane crashed into a field in Pennsylvania.

HUNTING A TERRORIST

In the weeks and months following these mass murders, U.S. military forces swung into action. They began the hunt for bin Laden and other al-Qaeda leaders. That search first took them to the Asian nation of Afghanistan. Al-Qaeda had maintained its main training bases there. The Americans came close to nabbing bin Laden, but he had managed to slip away.

As the hunt continued, more months and then years went by. Bin Laden benefited from a network of people who hated Americans as much as he did. They helped to keep him hidden from U.S., British, and other Western forces.

Eventually, however, bin Laden's luck ran out. Late in 2010, American spy organizations zeroed in on a large house in Afghanistan's neighbor, Pakistan. Satellite photos showed the house sitting in a sprawling compound ringed by high concrete walls. It appeared that a major terrorist might be concealed there. Further study narrowed down his identity. As one U.S. official says, the experts figured "there was a strong probability that the terrorist hiding there was Osama bin Laden."

By mid-February of 2011, President Barack Obama decided on "an aggressive course of action."

Osama bin Laden

U.S. President Barack Obama (*second from left*) and Vice President Joe Biden (*far left*), along with members of the national security team, receive an update on Operation Neptune's Spear, a mission against Osama bin Laden, in one of the conference rooms of the Situation Room of the White House on May 1, 2011.

Obama's BOLD MOVE

Bin Laden's compound in Abbottabad, Pakistan, was much larger and more fortified than any other homes in the area. Also, it was located less than a mile from Pakistan's Military Academy. These facts bothered President Obama and other high U.S. officials. It was hard to believe that no Pakistanis knew someone important was hiding in the compound. Obama suspected that Pakistani military leaders might know. So he decided not to notify the Pakistani government about the raid. This was an extremely bold move. Normally a nation does not send military forces into another's territory without permission. Many countries view such a move as an insult at the least. For that reason, many Pakistanis were angry about the U.S. operation.

RELENTLESS TRAINING

In a secret message, Obama ordered the Navy SEALs to begin training. Their mission was to sneak into Pakistan in the dead of night and kill bin Laden. The SEALs' training was no less secretive. Beginning in February 2011, they rehearsed every possible move they might make during the mission. A full-sized replica of the Pakistani compound was built. This added an extra element of realism to the training.

Meanwhile, the SEALs put themselves through punishing physical exercises and drills. They hardened their muscles and sharpened their senses. In April, the Navy secretly flew the commandos to Afghanistan. There, the relentless training continued. Finally, they were ready. Obama gave them the go-ahead and the mission, code named Operation Neptune Spear, began.

NEVER TO GIVE UP

The Navy SEALs who completed the mission are special, elite warriors. They receive longer and much more grueling training than ordinary soldiers do, so they are able to perform operations that are too difficult for regular soldiers.

The Navy Special Warfare Trident Insignia, worn by qualified U.S. Navy SEALs

The SEALs are only one of several U.S. military groups labeled Special Operations Forces (SOF). They are often called "special ops" groups for short. The U.S. Army has several such groups. One is the Army Special Operations Forces, better known as the Green Berets. Two others are the Army Rangers and Delta Force. The Marines have the Marine Corps Force Reconnaissance, or "Force Recon." And the Air Force and Coast Guard have special ops outfits of their own.

No single special ops group is more skilled than another. Rather, each has its area or areas of expertise. The SEALs are particularly skilled in operations in water settings, for example. And the Army Rangers and Air Force Pararescue teams are highly adept at jumping from planes. Thus, each special ops group is extremely good at what it does.

A SEAL student struggles to lift his portion of a heavy log over his head during Log PT (physical training) at the Naval Special Warfare Center.

All members of the U.S. special ops groups receive roughly the same basic training. Unbelievably tough, it molds them into what some experts have called the finest fighters in the world. It is not just their high degree of physical fitness. Nor is it just their equally excellent mental fitness. It is also a never-say-die attitude. One special ops soldier remarked that the training is about "never quitting, never giving up, regardless of the difficulty or pain."

Members of SEAL Team 4 participate in a training exercise.

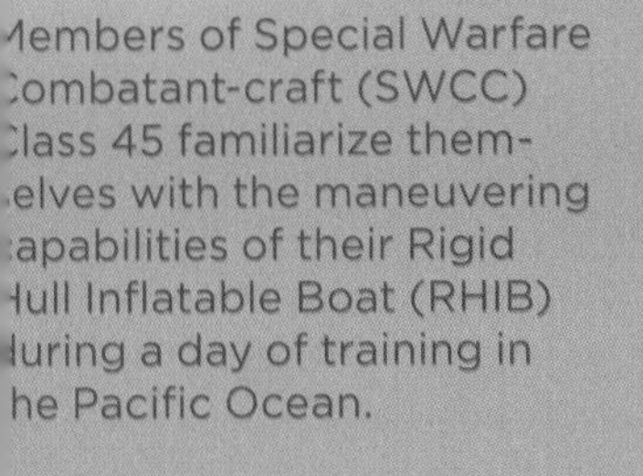

Members of Special Warfare Combatant-craft (SWCC) Class 45 familiarize themselves with the maneuvering capabilities of their Rigid Hull Inflatable Boat (RHIB) during a day of training in the Pacific Ocean.

fact BOX

The Special Boats Units

Besides the SEALs, the Navy has another special ops group. It is the Special Warfare Combatant-craft Crewmen, or SWCC. (It is pronounced "swick.") Another name for it is Special Boat Units. Its members are experts in the use of small boats for special missions. Often they carry units of SEALs into enemy lands.

CHAPTER TWO

A squad of SEAL Qualification Training candidates submerge themselves in near freezing water during a rewarming exercise in Kodiak, Alaska.

TRAINING the Body

Members of the various U.S. Special Operations Forces are often asked different versions of the same question. One version wants to know which special ops group is the best. Another asks if the Navy SEALs are tougher than the Army Rangers. Still another inquires whether the Marines in Force Recon are better than the soldiers in the Green Berets.

"I'll let you in on a little-known secret," former Air Force officer Rod Powers says. "There aren't any 'best.' It's like asking which is the best doctor, a brain surgeon or a heart surgeon?" Both of those doctors can "treat many illnesses," Powers continues. "However, each are 'best' in their specific specialties. Special Operations Forces are just like that. Each are highly trained . . . [and] can be used for many general special operations missions. However, each Special Operations Group is primarily trained for specific types [of] missions."

That training members of the U.S. special ops groups undergo can differ in kind and emphasis. For instance, the Navy's Special Boat Units get more detailed instruction in operating watercraft than the Army Rangers do. Also, the Air Force Pararescue units usually complete more parachute jumps in their careers than the Green Berets do.

However, all of them must first achieve a high level of physical conditioning. To that end, they all go through a period of basic physical training. Members of the various special ops groups do the same or similar basic exercises and drills. Moreover, their workouts all have roughly the same degree of difficulty.

fact BOX

Training at Fort Benning

Army Rangers get their basic training in the first of three training phases. It is called the Benning Phase. This is because it takes place in the Ranger school at Fort Benning, Georgia. The students train up to nineteen hours a day, seven days a week. Among other things, they do repeated physical exercises designed to increase their toughness and endurance.

An Army Ranger during the water confidence test event at the Best Ranger Competition at Fort Benning in Georgia

PUSHING THE LIMITS

That degree of difficulty is naturally very high. After all, special ops fighters are expected to be, and *are,* the best of the best. On their missions they often encounter life and death situations. And in such circumstances, being in top shape is a must.

Consider a case in which one man in a unit is not as physically fit as the others. He might not be able to perform his assigned tasks well enough. That could cause the mission to fail. Even worse, one or more members of the unit could die because of his lack of conditioning.

Special Operations Forces leaders want to avoid such situations. So the physical training they give their men pushes them to the very limits of human endurance. This often involves enduring considerable discomfort. A certain amount of pain is also expected. The basic training "will max you out physically in every exercise you attempt," says former Navy SEAL Stew Smith. "You can become tougher by working out harder," he explains. This approach gives the body "an increased ability to build its pain tolerances without getting yourself injured."

The Flexible SEALs

In the name Navy SEALs, the word SEAL stands for "Sea, Air, Land." This indicates that the members of the group are trained to fight in all three of these settings. So they are very flexible. And such flexibility makes them highly effective.

SEAL candidates "surfing," or laying in the surf as the frigid water rolls over them. Surfing is a tactic used to keep the candidates cold and wet and to weed out the people who don't have the perseverance to continue.

"THE MORE YOU SWEAT"

The basic physical exercises that make the recruits tougher are fairly standard among the various special ops groups. They include pushups, sit-ups, pull-ups, and long runs. Typical is the punishing exercise program for would-be Navy SEALs. They must perform several sets of exercises in a row with little or no rest between sets.

First, the trainees must do a minimum of forty-two push-ups in fewer than two minutes. Right after that they have to do fifty sit-ups, also in under two minutes. Then come six pull-ups. Finally, the already exhausted young men do a 1.5-mile run in under 11.5 minutes.

In a later phase of basic training, the SEAL trainees undergo the dreaded Hell Week. It consists of five days of seemingly endless exercises and tasks. During the entire time, the recruits are wet, cold, and hungry. Also, they get almost no sleep.

Naval Special Warfare SEAL Qualification Training candidates learn how to rappel from a sixty-foot tower at the Naval Special Warfare Center.

In the WATER ALL DAY

Tom Hawkins served as a Navy SEAL in the late 1960s. In the following passage, he recalls a few vivid snapshots of the training phase known as Hell Week.

> I remember almost everything about Hell Week, especially being wet, cold, and dirty. . . . We started out with about 100 people. And we graduated 28. . . . By the time we got down to the twenty-eight [guys], we made our [required seven-mile] swim. [But] with the currents and the tidal changes, by the time a person actually finished the swim, it was [more] like fifteen miles. We were in the water with no food, no fresh water, no nothing, for all day.

Members of Basic Underwater Demolition (BUD/S) /Sea, Air, and Land (SEAL) Class 244 endure the bitter cold while conducting exercises as part of their BUD/S confidence and personal endurance training at the Naval Amphibious Base in Coronado, California.

SEAL trainees during Hell Week

Other U.S. special ops groups have their own versions of Hell Week. For instance, the Army Rangers go through a four-day endurance test called Cole Range. And Air Force special ops candidates endure Motivation Week. One graduate of Motivation Week called it "ugly" and "downright evil." He added, "It's nonstop training, constant screaming . . . and only a couple hours of sleep a night. When you finally get a chance to put your head on a pillow at night, you're out in seconds."

Because special ops basic training is so hard, it has a huge failure, or drop-out, rate. Only 25 percent, or one in four, of those who start the training end up completing it. But this pays off later in real combat situations. Members of the special ops community have an old saying that sums it up well. It goes: "The more you sweat in training, the less you bleed in combat."

SEAL Qualification Training students endure a long hike after finishing their second day of close quarters combat instruction.

CHAPTER THREE

Preparing the MIND

Before beginning their basic training, all would-be special ops fighters know it will be physically hard. But few realize they will face a major mental strain as well. In truth, the mind is tested no less than the body during the grueling workouts. "At every stage" of special ops basic training, says Air Force Master Sgt. Pat McKenna, "your body and mind get taxed." Indeed, they are pushed to the "breaking point, to the very brink of total collapse. And you must find somewhere deep within yourself the grit to forge ahead." One must do this, he adds, "even when every sinew screams 'uncle.'"

Successful trainees find that "grit" by learning to sort through and control certain key feelings and emotions. Among them are stress, fear, and the will to keep going when things get difficult. If a recruit can master these feelings, odds are he will complete the training. However, the opposite is also true. Indeed, many military experts say that lack of mental preparation is the chief reason for trainees quitting the program.

DEALING WITH STRESS

Those same experts point out that it is perfectly normal to feel stress during basic training. After all, the recruits find their bodies pushed to their physical limits. The instructors yell at and berate them on a regular basis. Meanwhile, the young men must adhere to an incredibly tight schedule. And they get far less sleep than they are used to.

When under such pressures, it is only natural to feel some degree of mental strain or anxiety. The challenge for a trainee is how he deals with these normal stresses. Will they overwhelm him and prevent him from performing his assigned tasks? Experience shows that the successful trainees are the ones who don't allow this to happen.

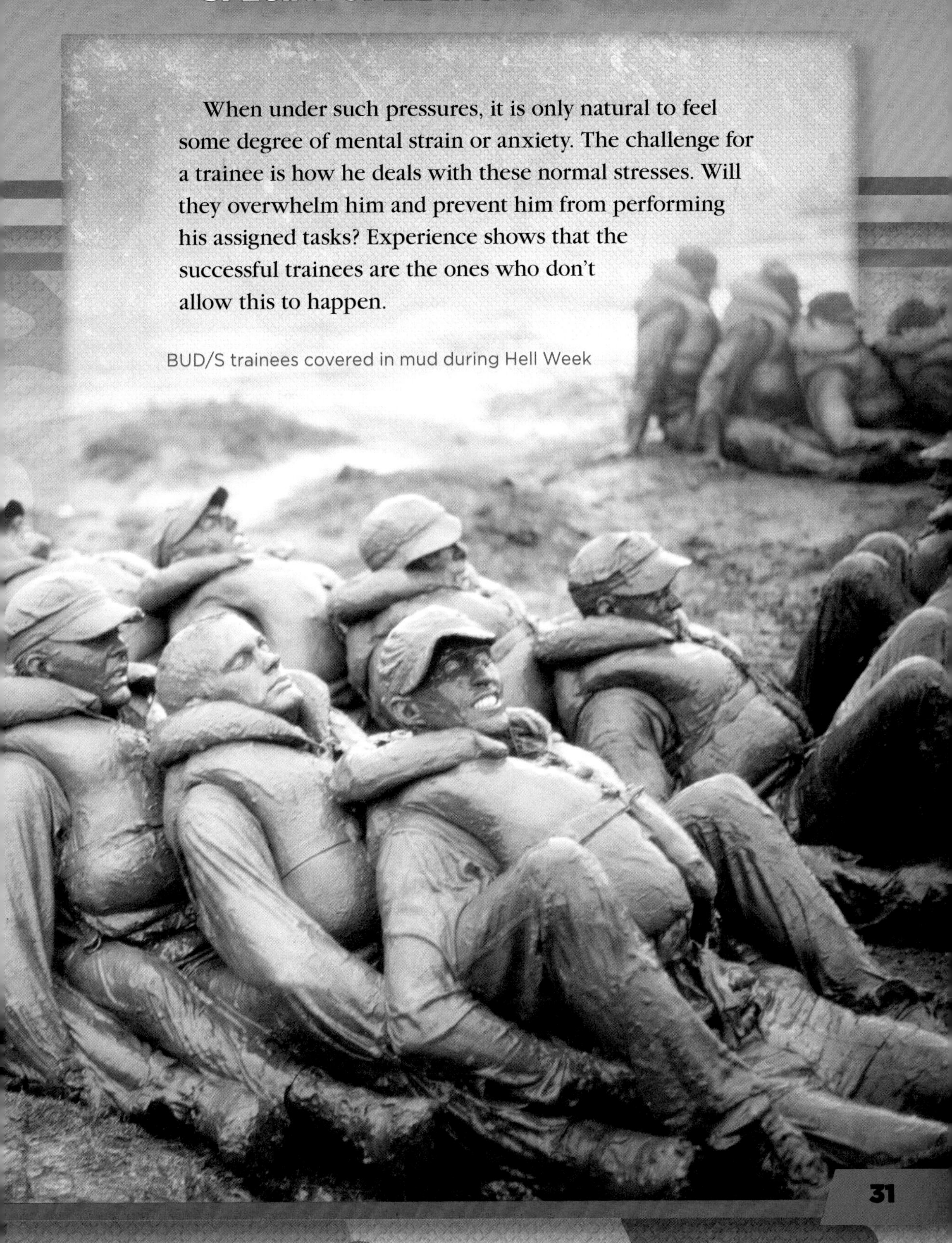

BUD/S trainees covered in mud during Hell Week

A member of Basic Underwater Demolition (BUD/S) /SEAL Class 244 navigates an obstacle on the training course at the Naval Amphibious Base in Coronado, California.

THE TOTAL PACKAGE

Some young men do learn to deal with the stress but then encounter other problems. For example, a number of trainees find it hard to accept criticism. They may interpret the instructors' gruffness and demanding tone as personal attacks on their character. A few recruits even become convinced that the trainers are out to get them. This can lead to feelings of mistrust, fear, and resentment.

But such feelings are unfounded. Any and all of the instructors' criticism is meant to be constructive. The trainers push the recruits hard—mentally as well as physically—for their own good. The instructors know that real combat is chaotic and scary and pushes fighters to their mental limits. Basic training is designed partly to prepare the young men for such situations. It "can never really simulate combat," former Air Force special ops fighter Jeremy Hardy admits. But what the instructors dish out "is very adept at preparing you for the real deal. The stress, both physical and mental," of the training "is as close as it gets."

Steve Sanko, a special ops trainer for the Army, also calls attention to the mental component of training. In his view, it's a matter of mind over muscle. "It doesn't make a difference if you're [a] big, buff stud," he says. "You've got to have smarts and a heart" along with physical strength and endurance. The trainees with the best chance are those with "the total package."

fact BOX

The Humor Factor

Another crucial mental factor in special ops training is having a sense of humor. Dwight D. Eisenhower, the great Army general who became a U.S. president, explained its importance in molding first-rate soldiers. "A sense of humor is part of the art of leadership," he said. It is essential in "getting things done."

U.S. Army Specialist Hospedales gives candy to an Iraqi girl during a patrol in Muqdadiyah, Iraq.

THE EDGE OF COMBAT SPORTS

To succeed in special ops basic training, recruits must deal with an array of mental factors. They have to learn to handle stress and pain. They must also be able to accept criticism without complaint and to overcome fear.

Dealing with these emotional aspects of the training is difficult for all trainees. But those with former experience in combat sports have at least a slight edge over those without it. Combat sports include wrestling, boxing, and martial arts. The discipline required for these activities can help prepare someone for the mental stresses of special ops training.

One person who benefited this way was a Navy officer, Captain Ryan McCombie, who served as a SEAL in the late 1980s. He had been a high school wrestler. And he later recalled how that experience helped him get through the SEALs basic training program.

As a wrestler, McCombie said, he had faced a sort of combat situation in each wrestling match. He knew in advance that he would endure fatigue and muscle pain in a bout. He also learned that he would have to be 100 percent focused on the task at hand, and this required considerable mental discipline.

McCombie also repeatedly faced the reality that he might lose a match. That created a certain amount of stress and fear. The only way to overcome these negative feelings, he discovered, was to be mentally prepared. Later, he applied this same approach to overcoming mental challenges in his SEAL training and missions.

MUSCLES Have MEMORY

Navy SEAL Ryan McCombie had been a high school wrestler. Successful high school, college, and Olympic wrestlers are known for their high degree of mental discipline. "The discipline of the sport stayed with me," he later said when describing his basic training for the SEAL program.

> My time in wrestling . . . had hardened me mentally, and muscles had memory. I was able to stand up to what the [SEAL school] instructors threw at us. It wasn't easy. I was tired and hurting most of the time. But I had become accustomed to that when I was wrestling.

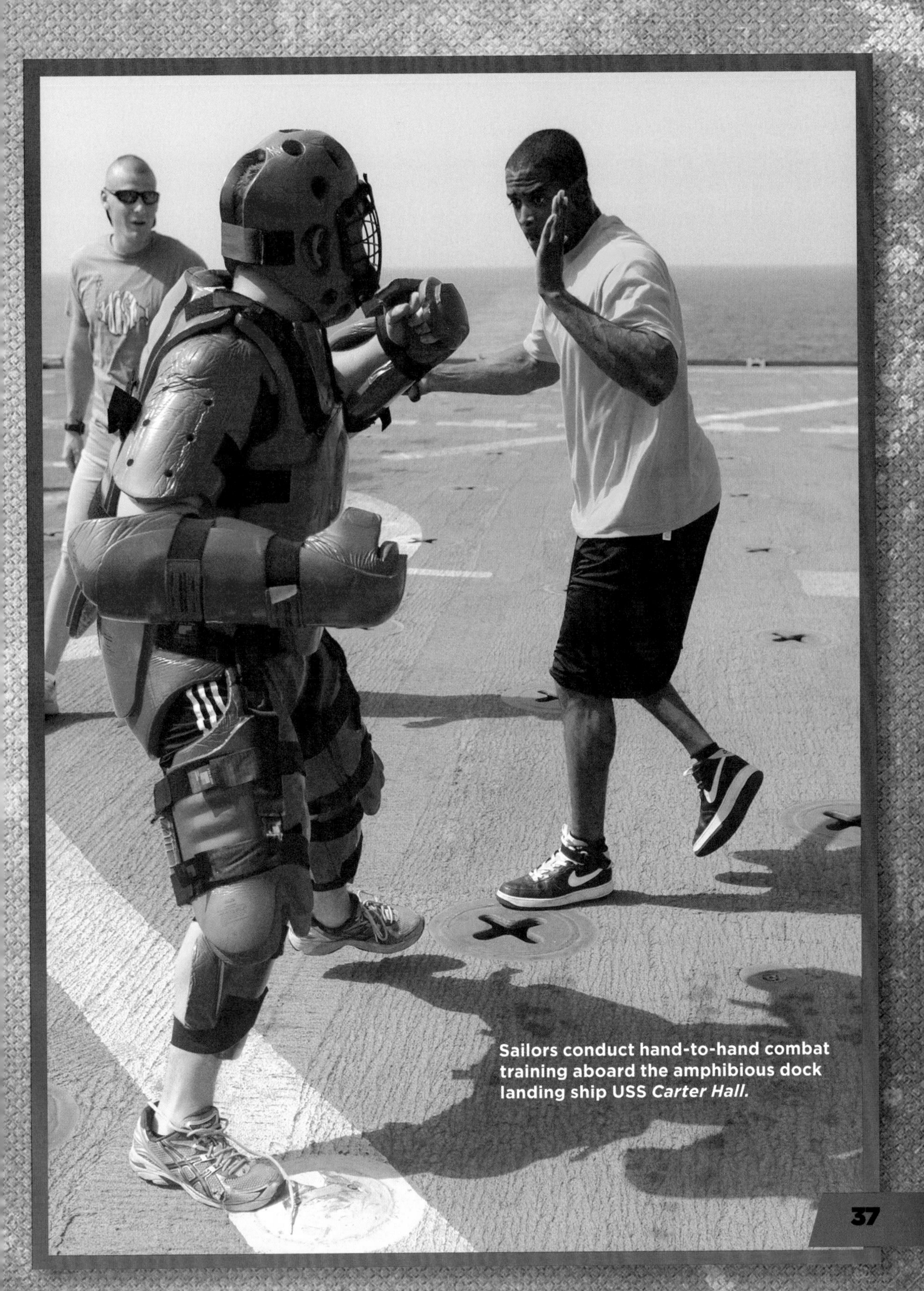

Sailors conduct hand-to-hand combat training aboard the amphibious dock landing ship USS *Carter Hall.*

CHAPTER FOUR

U.S. Marines of the 5th Air Naval Gunfire Liaison Company learn mountain warfare skills from the 2nd Republic of Korea Marine Division Rangers, Mountain Training instructors.

Lessons in SURVIVAL

Preparing the body and mind is only the initial phase of special ops training. The elite fighters in all of the U.S. military branches also learn survival methods. In their missions, they may find themselves in distant, deep forests. Or they may be thrust into arid desert settings or jungles crawling with poisonous snakes. Moreover, any of these locales may be inside enemy territory. The fighters have to be able to survive for an unknown length of time. They hope to be able to do it without being detected and captured by opposing forces.

BASIC SURVIVAL METHODS

Schools that teach special ops survival methods exist in various parts of the United States. The Air Force Survival School is at Fairchild Air Force Base near Spokane, Washington. Army survival training takes place near Fort Bragg, North Carolina. The Navy SEALs learn many of their survival techniques in the wilds of northern Maine.

These places and others like them emphasize a kind of training called SERE. The letters stand for Survival, Evasion, Resistance, Escape. To some degree, the training varies from one military branch to another. The SEALs' courses specialize in ways to survive in watery environments, for example. Pararescue recruits learn to make it home after their aircraft crashes. Meanwhile, Army Rangers become adept at surviving in mountainous terrain.

fact BOX

The Rangers in the Mountains

The "Mountain Phase" of Army Rangers' training occurs near Camp Merrill, in Georgia. It lasts twenty-one days. The students learn basic climbing techniques and how to rappel down a rock face. They also learn to build makeshift shelters and eat off the land.

An Army Ranger instructor explains the technical instructions of rappelling from the fifty-foot rock to his left in Dahlonega, Georgia.

Students from BUD/S Class 284 participate in a land navigation training exercise.

All special ops recruits learn some basic, general survival methods. For example, all must pass exhausting swimming tests. This is because on a mission any of them might end up in a river, a lake, or the ocean. They also learn how to find food and drinking water in diverse wilderness settings.

In addition, the recruits are taught how to build fires and how to give first aid to both themselves and wounded comrades. And they all receive some form of instruction in navigation. This includes map-reading and locating north and other compass directions. It also includes finding one's way out of remote or hostile areas. Harry Haug, a Navy SERE instructor in Maine, sums up these universal basic survival methods. "We teach primitive means of making due with what is at arms-reach," he says. Among these techniques are "constructing a fire with flint [rock] and steel," and recognizing "what's edible" in a given area. Haug and the other instructors also teach "how to use a simple piece of metal as a compass."

Weighing the NEED for a FIRE

The U.S. military has issued several manuals listing survival methods. One of the best is the *Army Ranger Handbook*. It tells how to find one's way through mountains and forests. There are also instructions on how to make a fire. Included in that section is a warning about the pitfalls of building a campfire in enemy territory. It "creates smoke," the manual says, "which can be smelled and seen from a long distance. It [also] leaves signs of your presence. Remember, you should always weigh your need for a fire against your need to avoid enemy detection."

DEALING WITH THE ENEMY

Another important aspect of special ops survival training is how to deal with enemy forces. The recruits first learn to evade capture by those forces. In part, this involves mastering the art of camouflage, or masking oneself to stay hidden. One approach is to wear clothing that helps the wearer blend with the natural surroundings. The recruits also learn to camouflage their makeshift shelters, called "hides."

The training addresses other aspects of dealing with an enemy as well. One of these is learning to track opposing soldiers. Another is how to resist capture. Still another is what to do if one is actually captured.

A Combat Engineering Sniper in a camouflage suit

WAR GAMES

Learning such things is important. Equally vital is putting them into practice. So near the end of the training the recruits are released into a wilderness setting. There, they must deal with an "enemy" army made up of special ops instructors.

The recruits do their best to elude the enemy. But as a popular special ops Web site says, "after several days of the war game . . . the students are captured. [They] are then taken to a mock prisoner of war camp." There, "the 'prisoners' are hooded and roped together." They also endure days without food or sleep. This treatment is "severe enough that, over the course of survival school, a student typically drops 15 pounds."

Creating such harsh conditions mimics real life. That makes the experience very valuable for the recruits. In fact, overall, special ops survival training is highly effective. Army special forces officer Michael Durant can testify to that. In 1993 his helicopter crashed during a mission in the African nation of Somalia. The incident inspired the famous movie *Black Hawk Down.* Durant was captured and suffered abuse at the hands of the enemy, but he eventually made it home. He credits his forceful survival training with saving his life.

UH-60 Black Hawk helicopter

fact BOX

Authentic Enemy Captors

In the mock battle the special ops recruits undergo, the "enemy" fighters are their instructors. After capturing the students, the instructors look and act just like foreign enemies. Often they even speak a foreign language to keep the recruits in a state of confusion. This approach makes the capture phase of the training very authentic.

U.S. Marines assigned to the 3rd Marine Regiment role play as opposition force soldiers captured and held as enemy prisoners of war during a simulated night raid at Kahuku Training Area in Oahu, Hawaii.

Army Rangers from the 1st Battalion, 75th Ranger Regiment patrol the streets of a mock city before evacuating in an MH-47 Chinook during a capabilities training exercise conducted at Fort Bragg in North Carolina.

CHAPTER FIVE

Experts in WEAPONS Use

U.S. special ops fighters are experts with numerous deadly weapons. These range from rifles, handguns, and other handheld firearms to missiles and other larger-scale killing machines. Instruction in the use of these weapons is a long-term process. It begins in basic training. There, the special ops recruits learn to use a few standard military arms. Later, the weapons instruction intensifies. The trainees become experts with several other weapons that ordinary soldiers never have access to. A highly trained member of special ops in a sense becomes one with his weapons. He can even take apart and reassemble most of them blindfolded.

M4 carbine rifle

THE MOST COMMON COMBAT RIFLE

The first rifle that special ops trainees learn to us is the M4 carbine. It is the most common combat rifle in modern warfare. The M4 shoots between 700 and 950 rounds, or bullets, per minute. So it can cause a lot of damage to people and property. Therefore it is often used in urban, or city, combat. The M4 is fairly light for such a destructive weapon. It weighs only about 7 pounds (around 3 kg). That makes it easy to carry.

The M4 has long been and remains an effective rifle. But some special ops soldiers started complaining about it in 2001-2002. That was when U.S. forces began fighting in Afghanistan. The terrain there is very rugged and features many sandy deserts and plains. "Even with the dust cover closed," one special ops soldier said, "sand gets all inside." The rifle "still fires, but after a while the sand works its way all through the gun and jams start."

HK416 rifle

Rumors say that this problem made members of the elite Army Delta Force particularly unhappy. Supposedly they requested that a new rifle be made. It is called the HK416. ("HK" stands for Heckler and Koch, the German arms company that developed the rifle.) Basically an upgraded version of the M4, it is said to be far less likely to jam because of sand.

Other rumors say that some SEAL units have also adopted the HK416. In fact, a reliable military source asserts that an HK416 fired the bullet that killed Osama bin Laden in 2011. "The stack of SEAL assaulters [who] went through Osama bin Laden's bedroom door were [holding] HK416s," the source claims. The military has not confirmed this. As might be expected, it remains tight-lipped about that highly secret mission.

Soldiers from the 101st Airborne Division conduct a raid in Samarra, Iraq.

OTHER LETHAL WEAPONS

Another weapon that special ops fighters commonly train with is the MP5. It is a submachine gun often employed in urban warfare. The MP5 is an example of an automatic weapon. An ordinary gun fires a single round when the operator squeezes the trigger. In contrast, an automatic weapon fires a burst of bullets at one touch of the trigger.

MP5 submachine gun

A number of special ops fighters also like another automatic weapon—the M249. One of its benefits is that it uses the same ammo (short for ammunition) as the M4. So a fighter can switch from one gun to the other and needs to carry only one kind of ammo.

M249 light machine gun

Dummy hand grenades used for practice

All members of U.S. special ops groups are trained to use another common weapon, the grenade. It is essentially a small bomb. Some grenades scatter tiny pieces of metal when they explode. Others release smoke or tear gas to confuse or hinder an enemy. Trainees are also taught to use smoke grenades as signaling devices.

Many special ops trainees also learn to use larger and/or more specialized weapons. One is the M24 sniper rifle. It has a powerful telescopic site that allows the shooter to hit very distant targets. Training in the use of bigger and heavier antitank missiles is also available.

M24 rifle

fact BOX

Specs for the M24

The M24 sniper rifle weighs just over 15 pounds (6.8 kg). It has an effective range of 1,500 feet (457 m), or five football fields. Also, the rifle fires bullets at the unusually high speed of 2,571 feet (764 m) per second.

Special ops trainees learn to use many other weapons as well. Information about the arms a specific unit has used in a given secret mission is generally classified. But one thing is certain: These fighters receive very extensive weapons training, far more than any ordinary soldier.

That expertise in using lethal weapons is but one aspect of the U.S. special ops fighter's training. He combines it with his punishing physical and mental preparation. Added to these is the exhausting survival training he undergoes. Because all these first-rate training factors come together within him, he is arguably the finest and most lethal warrior in human history. For that reason, he and his comrades carry out the country's most dangerous missions.

The RELIABLE M9

One of the most effective and reliable weapons in the U.S. arsenal is a handgun. It is called the M9 Beretta. One of its advantages is that it is small, so it can be easily concealed. In addition, the M9 can be used in a number of extreme conditions. The U.S. military put it through a battery of rigorous tests. Former Air Force sergeant Rod Powers summarizes them:

M9 Beretta

> An extreme climatic test checked its ability to withstand temperatures between minus 40 and 140 degrees. A 10-day salt water immersion and humidity trial tested its resistance to corrosion. It tackled mud, sand, dirt and water to test its operation under adverse field conditions . . . and [in all cases] fired again with no problems.

A U.S. Army sniper team scans the horizon after reports of suspicious activity along the hilltops near Dur Baba, Afghanistan.

Source Notes

Chapter 1: Training to Kill Bin Laden

p. 10, "Today, we should all . . ." Elizabeth Stuart, "Navy SEALs Who Took Down Bin Laden Trained Hard," *Deseret News*, May 3, 2011, http://www.deseretnews.com/article/700132327/Navy-SEALs-who-took-down-bin-Laden-trained-hard.html.

p. 12, "there was a strong . . ." Mark McGivern, "Osama Bin Laden: Revealed: U.S. Special Forces Trained for Weeks for 'Kill Operation,'" *Daily Record*, May 3, 2011, http://www.dailyrecord.co.uk/news/uk-world-news/2011/05/03/osama-bin-laden-revealed-us-special-forces-trained-for-weeks-for-kill-operation-86908-23103774/.

p. 16, "never quitting . . ." Bill Faucett, *Hunters and Shooters: An Oral History of the U.S. Navy SEALs in Vietnam* (New York: William Morrow, 1995), 264.

Chapter 2: Training the Body

p. 20, "I'll let you . . ." Rod Powers, "U.S. Military Special Operations Forces," http://usmilitary.about.com/od/jointservices/a/specialops.htm.

p. 22, "will max you out . . ." "Stew Smith Fitness," http://www.stewsmith.com/linkpages/toughestspecops.htm.

p. 26, "I remember . . ." Faucett, *Hunters and Shooters,* 145-148.

p. 27, "ugly . . . out in seconds," Master Sgt. Pat Mckenna, "Superman School," http://www.militaryspot.com/career/featured_military_jobs/.

p. 27, "The more you sweat . . ." Ibid.

Chapter 3: Preparing the Mind

p. 30, "At every stage . . ." McKenna, "Superman School."

p. 33, "can never really . . ." "Join the Air Force: Pararescue Apprientice," PopularMilitary.com, http://www.popularmilitary.com/jobs/pararescue.htm.

p. 33, "It doesn't make . . ." Ibid.

p. 34, "A sense of humor . . ." "Quotations By Subject: Humor," http://www.quotationspage.com/subjects/humor/.

p. 36, "The discipline of the sport . . ." Faucett, *Hunters and Shooters*, 257, 262.

Chapter 4: Lessons in Survival

p. 42., "We teach primitive . . ." "Navy SERE Training: Learning to Return with Honor," Navy News Service, http://usmilitary.about.com/od/navytrng/a/sere.htm.

p. 43, "creates smoke . . ." *U.S. Army Ranger Handbook* (Fort Benning, GA: United States Army Infantry School, 2000), 11-19.

p. 45, "after several days . . ." "Survival Training," U.S. Army Special Forces, http://www.training.sfahq.com/survival_training.htm.

Chapter 5: Experts in Weapons Use

p. 50, "Even with the dust . . ." "The USA's M4 Carbine Controversy," *Defense Industry Daily*, November 21, 2011, http://www.defenseindustrydaily.com/the-usas-m4-carbine-controversy-03289/.

p. 51, "The stack of SEAL assaulters . . ." Paul Bedard, "The Gun that Killed Osama Bin Laden Revealed," *US News*, May 11, 2011, http://www.usnews.com/news/washington-whispers/articles/2011/05/11/the-gun-that-killed-osama-bin-laden-revealed.

p. 55, "An extreme climatic test . . . " Rod Powers, "United States Military Weapons of War," http://usmilitary.about.com/od/armyweapons/l/aainfantry2.htm.

Glossary

ammo: Short for ammunition, the bullets used in various firearms.

automatic weapon: A machine gun or other weapon that fires a burst of bullets when the shooter squeezes the trigger.

barracks: On a military base, the dorm-like buildings in which recruits sleep.

beret: A small cloth cap worn by members of the Army Special Forces.

camouflage: Patterns and colors designed to make military uniforms, gear, and weapons blend in with a given natural setting.

commando: An elite, specially trained soldier who is assigned to difficult, dangerous missions.

compound: A group of related buildings, often surrounded by a fence or wall.

environment: A specific natural setting.

grueling: Very taxing or difficult.

hide: A makeshift, often camouflaged shelter erected in the wilderness.

hostage: A person who is held somewhere against his or her will.

navigation: Methods of finding one's location and plotting the way from one place to another.

reconnaissance: Scouting or investigating.

recruit: A soldier or other fighter who is in training.

round: An individual bullet fired by a gun.

special ops: Short for Special Operations Forces, consisting of the U.S. military's elite units of soldiers, like the Navy SEALs and Army Delta Force.

Bibliography

Byrne, John A. "My Story: From an Army Ranger in Iraq to Harvard." http://poetsandquants.com/2010/08/18/my-story-from-an-army-ranger-in-iraq-to-harvard/.

CNN. "How U.S. Forces Killed Osama bin Laden." http://articles.cnn.com/2011-05-02/world/bin.laden.raid_1_bin-compound-terrorist-attacks?_s=PM:WORLD.

Cooke, Tim. *U.S. Army Special Forces.* New York: Powerkids Press, 2012.

Couch, Dick. *Chosen Soldier: The Making of a Special Forces Warrior.* New York: Three Rivers Press, 2008.

De Lisle, Mark. *Special Ops Fitness Training: High-Intensity Workouts of Navy SEALs, Delta Force, Marine Force Recon, and Army Rangers.* Berkeley, CA: Ulysses Press, 2008.

Kyle, Chris et al. *American Sniper: The Autobiography of the Most Lethal Sniper in U.S. Military History.* New York: William Morrow, 2012.

Labrecque, Ellen. *Special Forces.* Mankrato, MN: Heinemann-Raintree, 2012.

Lunis, Natalie. *The Takedown of Osama bin Laden.* New York: Bearport, 2012.

Mann, Don. *Inside SEAL Team Six: My Life and Missions with America's Elite Warriors.* New York: Little Brown, 2011.

Montana, Jack. *Elite Forces Selection.* Broomall, PA: Mason Crest, 2011.

———. *Navy SEALs.* Broomall, PA: Mason Crest, 2011.

Nagle, Jeanne. *Delta Force.* New York: Gareth Stevens, 2012.

Nelson, Drew. *Green Berets.* New York: Gareth Stevens, 2012.

Roberts, Jeremy. *U.S. Army Special Operations Forces.* Minneapolis: Lerner, 2005.

Sandler, Michael. *Army Rangers in Action.* New York: Bearport, 2008.

———. *Pararescumen in Action.* New York: Bearport, 2008.

Sherwood, Ben. "Lessons in Survival: The Science that Explains Why Elite Military Forces Bounce Back Faster than the Rest of Us." http://www.thedailybeast.com/newsweek/2009/02/13/lessons-in-survival.html.

U.S. Navy. *The U.S. Navy SEAL Guide to Fitness and Nutrition.* New York: Skyhorse, 2007.

Von Wormer, Nicholas. *The Ultimate Air Force Basic Training Guidebook.* El Dorado Hills, CA: Savas Beatie, 2010.

Web sites

Army Enhanced Night Vision Goggles
http://www.army.mil/article/18980/army-fielding-enhanced-night-vision-goggles/

Army Special Forces Center
http://www.military.com/army-special-forces/training.html

Become a Combat Controller
http://usafcca.org/cca/careers/

How Apache Helicopters Work
http://science.howstuffworks.com/apache-helicopter.htm

Marine Force Recon
http://forcerecon.americanspecialops.com/

Official Web site of the Navy SEALs and SWCC
http://www.sealswcc.com/

Official Web site of SEAL Team Six
http://sealteamsix.net/

U.S.A.F. Pararescue. "Superman School."
http://www.pararescue.com/unitinfo.aspx?id=490

U.S. Army Special Forces. "Survival Training."
http://www.training.sfahq.com/survival_training.htm

U.S. Marine Corps Forces Special Operations Command
http://www.marines.mil/unit/marsoc/Pages/default.aspx

Weapons of the Special Forces
http://www.popularmechanics.com/technology/military/1281576

Index

Photo Credits

All images used in this book that are not in the public domain are credited in the listing that follows:

Credits:

Cover: Courtesy of the United States Navy
6-7: Courtesy of the United States Navy
8-9: Courtesy of the United States Navy
10: Courtesy of the United States Air Force
11 bottom: Courtesy of the United States Federal Bureau of Investigation
12: Courtesy of the United States Federal Bureau of Investigation
13: Courtesy of Pete Souza, Official White House Photographer
14: Courtesy of the United States Navy
15: Courtesy of the United States Navy
16: Courtesy of the United States Navy
17: Courtesy of the United States Navy
18-19: Courtesy of the United States Navy
21: Courtesy of the United States Army
23: Courtesy of the United States Navy
24-25: Courtesy of the United States Navy
26: Courtesy of the United States Navy
27: Courtesy of the United States Navy
28-29: Courtesy of the United States Navy
30-31: Courtesy of the United States Navy
32: Courtesy of the United States Navy
34: Courtesy of the United States Army
37: Courtesy of the United States Navy
38-39: Courtesy of the United States Marine Corps
40-41: Courtesy of the United States Air Force
42: Courtesy of the United States Navy
43: Courtesy of Dirk Beyer
44: Courtesy of Michael Shvadron, IDF Spokesperson Unit
45: Courtesy of jamesdale10
46-47: Courtesy of the United States Navy
48-49: Courtesy of the United States Army
50: Courtesy of the United States Army
51 top: Courtesy of Anynobody
51 bottom: Courtesy of the United States Navy
52 top: Courtesy of Dragunova
52 bottom: Courtesy of the United States Army
53 top: Courtesy of the United States Navy
53 bottom: Courtesy of the United States Army
55: Courtesy of the United States Army
56-57: Courtesy of the United States Army